The Girl and the Princess

by Dr Boaz

Illustrated by Chang Yit Wah

Includes a Guide to Problem-solving by Yeap Ban Har

WS Education

NEW JERSEY · LONDON · SINGAPORE · BEIJING · SHANGHAI · HONG KONG · TAIPEI · CHENNAI · TOKYO

Published by
WS Education, an imprint of
World Scientific Publishing Co. Pte. Ltd.
5 Toh Tuck Link, Singapore 596224
USA office: 27 Warren Street, Suite 401-402, Hackensack, NJ 07601
UK office: 57 Shelton Street, Covent Garden, London WC2H 9HE

National Library Board, Singapore Cataloguing in Publication Data
Name(s): Boaz, Dr. | Chang, Yit Wah, illustrator.
Title: The girl and the princess / by Dr Boaz ; illustrated by Chang Yit Wah.
Other Title(s): I'm a math star.
Description: Singapore : WS Education, [2022] |
          Includes a guide to problem-solving by Yeap Ban Har.
Identifier(s): ISBN 978-981-12-5058-3 (hardcover) |
          978-981-12-5112-2 (paperback) | 978-981-12-5059-0 (ebook for institutions) |
          978-981-12-5060-6 (ebook for individuals)
Subject(s): LCSH: Mathematical recreations. | Mathematics.
Classification: DDC 793.74--dc23

British Library Cataloguing-in-Publication Data
A catalogue record for this book is available from the British Library.

Copyright © 2022 by World Scientific Publishing Co. Pte. Ltd.

For any available supplementary material, please visit
https://www.worldscientific.com/worldscibooks/10.1142/12676#t=suppl

Desk Editor: Daniele Lee
Design: Eliz Ong

Printed in Singapore

It was a cool autumn day.

The girl was content to enjoy the breeze.

She turned to stretch

and was surprised to see

a young man pacing back and forth

under a tree a short distance away.

He looked very distressed.

The girl went up to the young man
and asked politely, "Are you lost, sir?"
The young man was startled,
and then he said, "Oh no, little girl,
I'm heading towards that castle.
But actually, I am feeling lost,
as I have a problem I don't know
how to solve."

The girl asked politely again,
"Could I try to help you
solve your problem, sir?"

The young man replied,
"The princess whom I love is held in that castle
under a curse from a wicked witch.

I doubt you can help me.

I'm a prince and the curse is beyond me

to break and you are but a little girl."

"Two are better than one," urged the girl.

"At least, let me try to help."

The prince was sceptical

but the girl looked sincere and bright

and ... he had nothing to lose.

So with a deep breath, he continued,

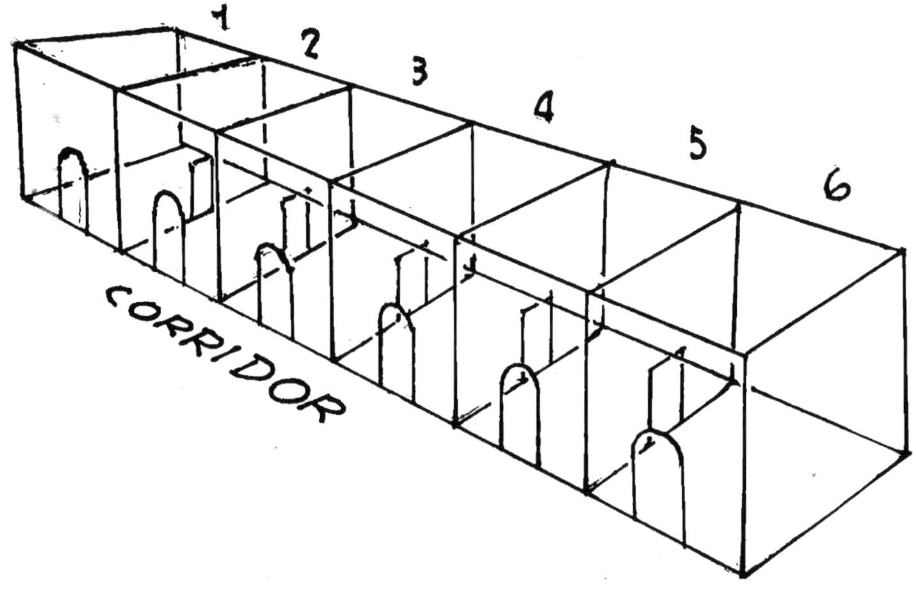

"The witch has imprisoned the princess in
a row of six rooms.

Each room has a door

to the two rooms at its sides,

except the end rooms,

which have but one door at its side.

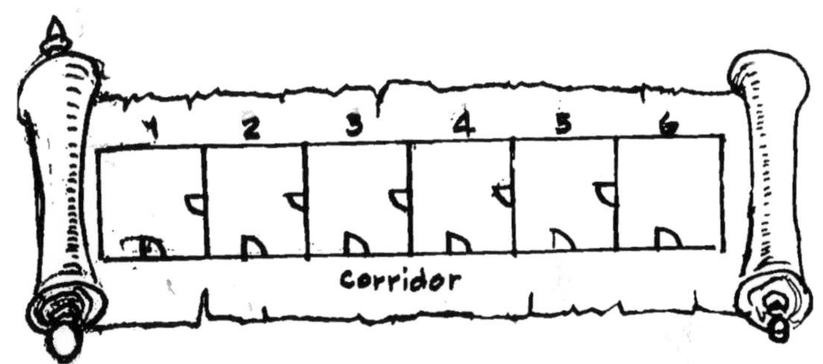

Now, the princess is only allowed to

stay in one room every day.

But the next morning at nine,

she has to move to an adjoining room.

If she breaks the rule, she will be

frozen for a hundred years.

All the rooms have a door to a

common corridor.

And the only way to break the curse is to knock

on one of these doors at noon.

If the princess is in that room when the door is

knocked,

the curse is lifted

and the door will open,

and I will rescue her!

However only one door is allowed

to be knocked on per day.

Break this rule and

the princess will be frozen

under the witch's curse.

Also, only eight knocks are allowed in total.

Nine days from today, when the sun rises,

and if I have not rescued her by then,

she will be frozen for a hundred years."

The girl gasped, then asked, "What is your plan to rescue the princess?"

"Well," the prince replied, "I was thinking of knocking on Door 3—it's somewhat in the middle—every day for eight days.

There is a good chance that the princess will be in Room 3 one of those days!"

The girl looked uncomfortable.

"My mother told me never to leave

important things to chance!

What if the princess moves like this—

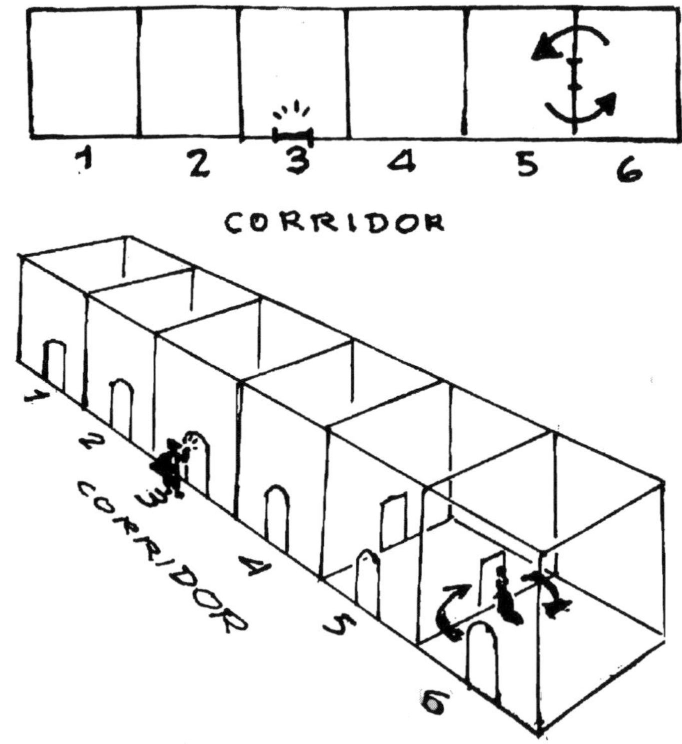

left and right between Rooms 5 and 6?"

"You are probably right," the prince
conceded despairingly.
"Here's another idea—
I'll knock from Door 1 and move on
to Doors 2, 3 and so on to Door 6.
Then I'll turn back to Doors 5 and 4."

The girl looked apprehensive.

"What if the princess starts at Room 2

and moves back and forth between

Rooms 2 and 1?"

The prince realised the danger and
shuddered at the consequence.

"We need a plan that will work whatever happens, and we have to consider all the possible movements of the princess.

A foolproof plan that does not depend on luck and coincidence,"
said the little girl firmly.

A refreshing gust of wind blew,

and the autumn leaves

playfully

drifted ...

The girl said softly, "It's usually good to
work on something simpler.
Let's try a row with only one room."

"That's too simple,"
said the prince, unconvinced.
"With just one room,
the princess will surely
be inside when I knock."

"That's true," the girl said.

"Let's try two rooms then.

If you knock on Door 1

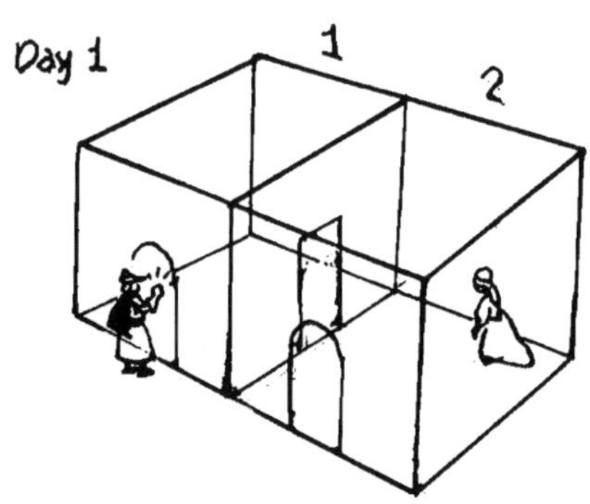

then on Door 1 again the next day,

you will find the princess!"

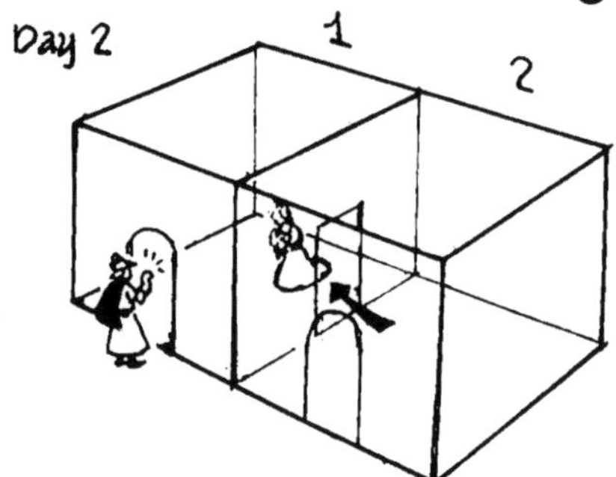

"Let's try three doors,"

the prince said eagerly.

"When there are three doors,

let's try knocking on the centre door,

which is Door 2, the first day,

and again on the next day.

Would that work?"

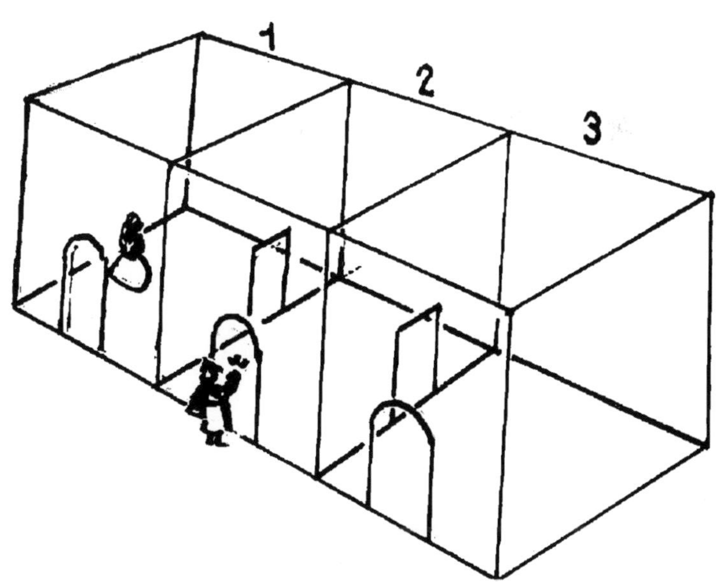

They checked by using a stick

to write on a sandy patch.

"Let's represent the situation by a grid,"

suggested the girl.

"The columns will be the room numbers,

and the rows will be the days."

The girl drew a fist to represent the door the prince would knock on, and she drew a crown for the princess.

If the princess started in Room 1, then the prince would rescue her on Day 2.

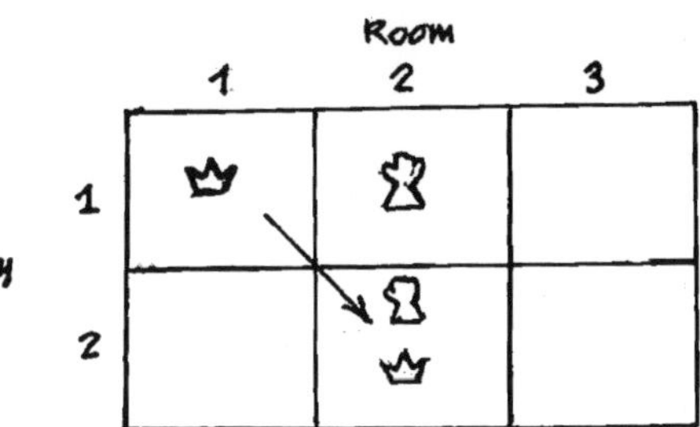

If the princess started in Room 2, then the prince would rescue her on Day 1 (but of course!).

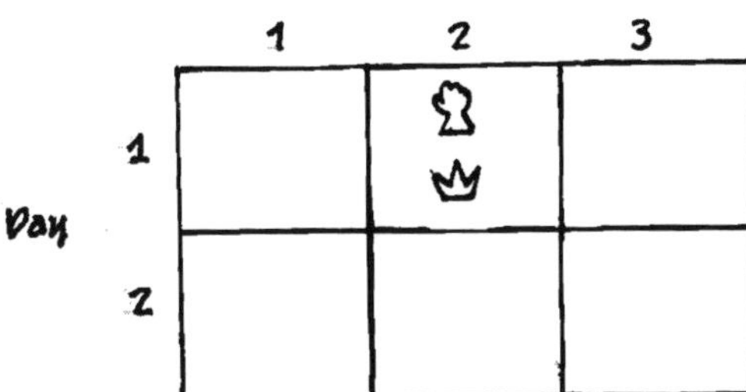

If the princess started in Room 3,

then the prince would rescue

her on Day 2.

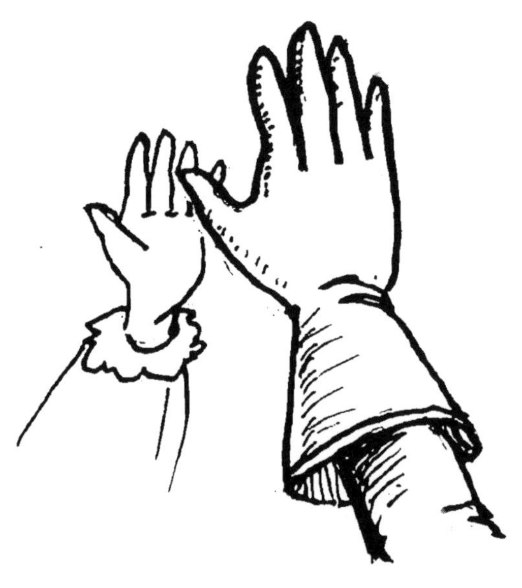

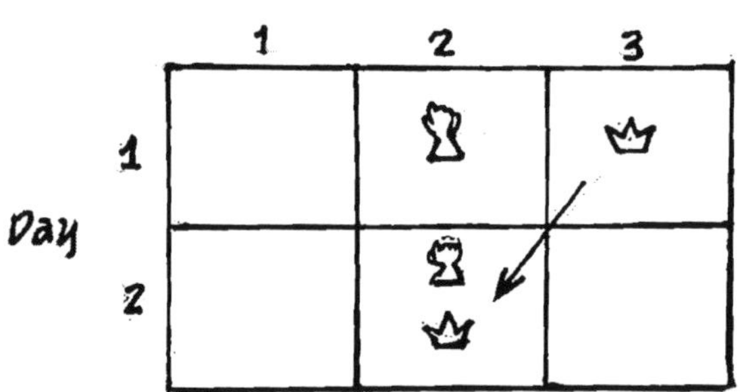

"Now, let's try four doors!" the prince said, growing excited.

This seemed more difficult.

"How about 2-2-3-3?" the prince offered.

They checked.

It would work if the princess started in Room 1.

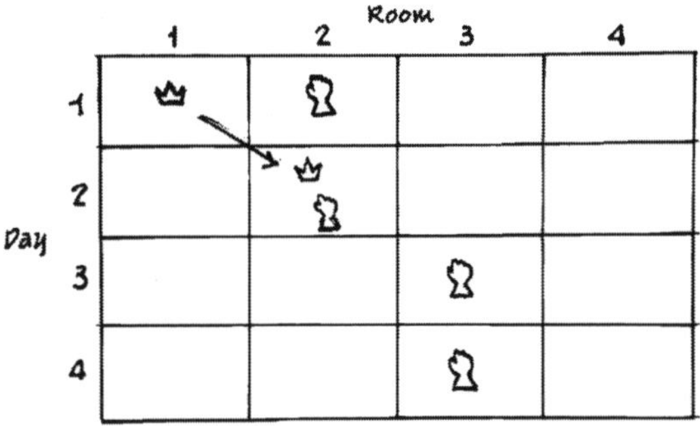

It would also work if the princess started in Room 2 (but of course!)

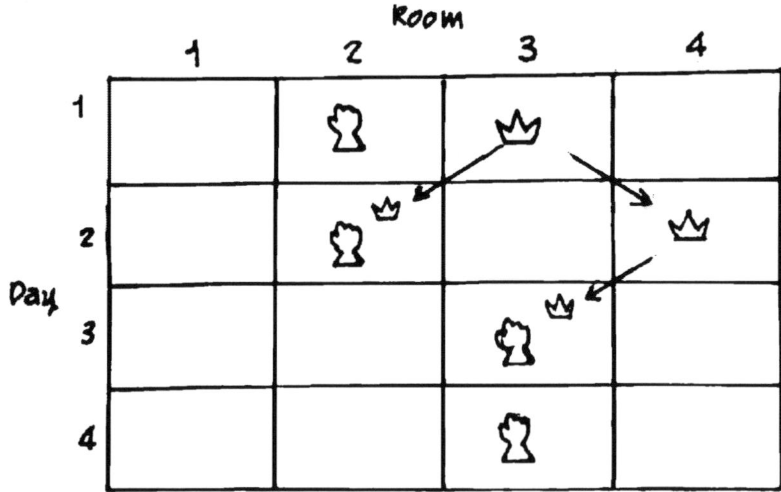

It would also work if the princess started in Room 3.

But, if the princess started in Room 4,

it would not work!

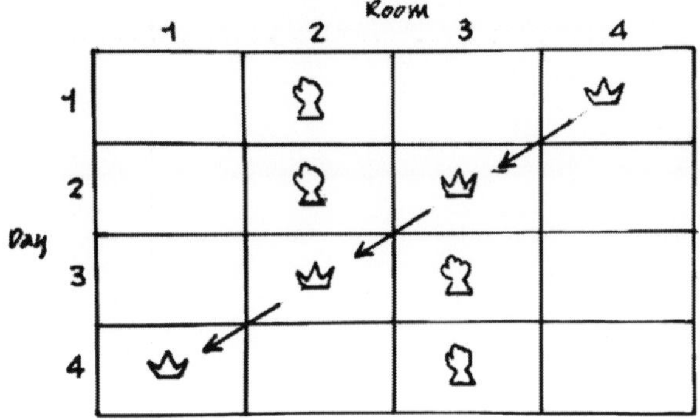

"Maybe it would work if we added

one more day," the girl guessed.

"Look!" and she added a fist to

Room 2 on Day 5.

And they were happy again.

And they pressed on with five rooms.

More leaves fell and the wind turned cold ...

Their minds froze.

"We need a new way to represent the situation," the girl said.

"Let's use the chessboard as the grid."

The girl fetched her chessboard,

a queen and some pawns.

"We'll just need the first six

columns of the chessboard.

These represent the six rooms.

The Queen chess piece

represents the princess.

Let's see how the princess moves.

If she starts from a white square on Day 1,

she will go left or right downwards

but will end up on a white square on Day 2.

On the third day,

she will go left or right

downwards again and

still end up on a white square!

The princess moves diagonally

on our chessboard representation!

If she starts on a white square,

she will always be on a white square.

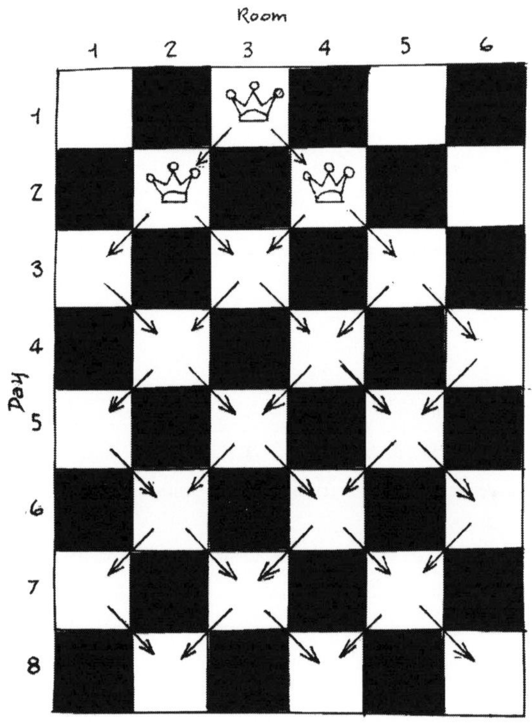

If she starts on a black square,

she will always be on a black square."

The girl stared intently at the board. "How do I rescue the princess if she starts on a white square?"

She said to the prince,

"The 8 pawns will represent which door you will knock on for each of the 8 days.

Let's start by building a diagonal wall on the white squares, across the 4 centre rooms."

And she placed the four pawns in a diagonal line—representing that the prince would knock on doors 5, 4, 3 and 2, one door for each of the first 4 days.

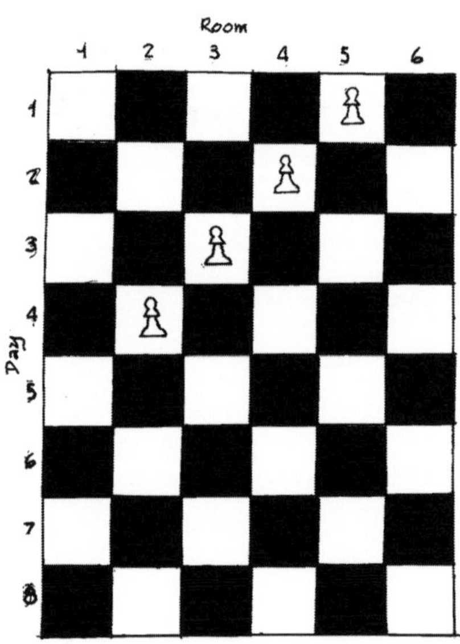

"If the princess starts on a white square, she has to move along white squares. Soon she will encounter a pawn which is placed on a white square."

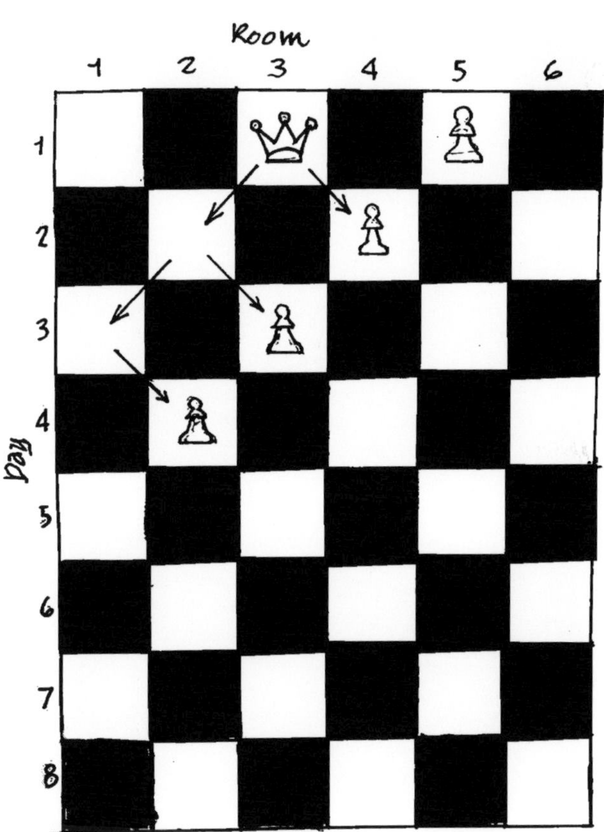

The girl smiled at the prince.
"Within four days,
you will rescue the princess!"

"However ... if the princess starts

on a black square,

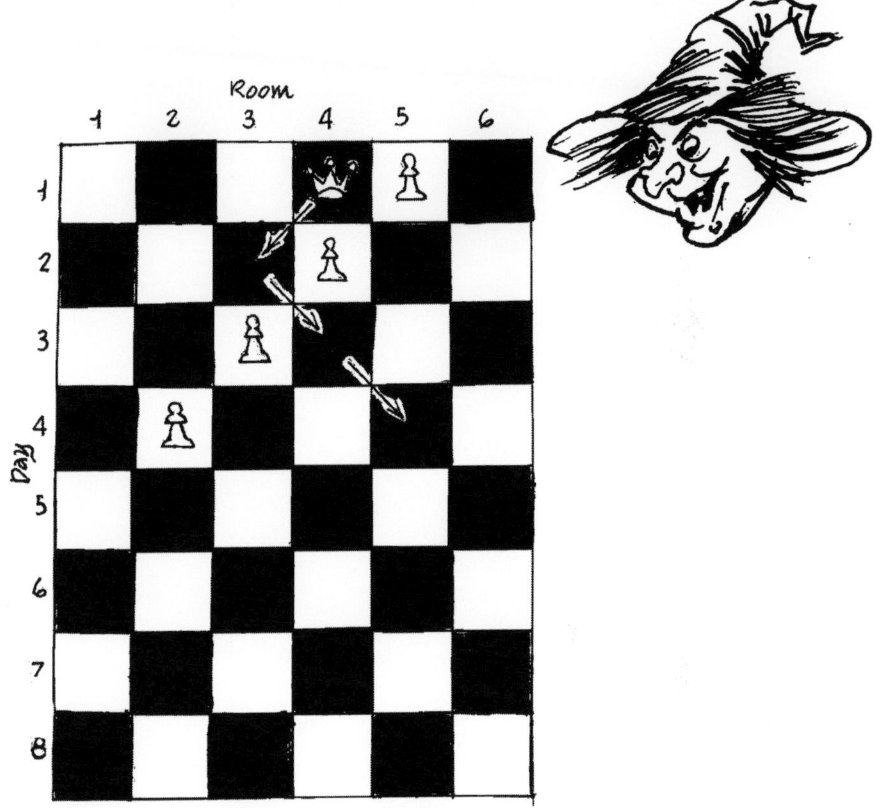

she will sadly avoid the wall of pawns

on the white squares!"

"This will be like knocking on Door 4
on one day while the princess is in Room 3
and then knocking on Door 3 the next day
while the princess has moved to Room 4!"

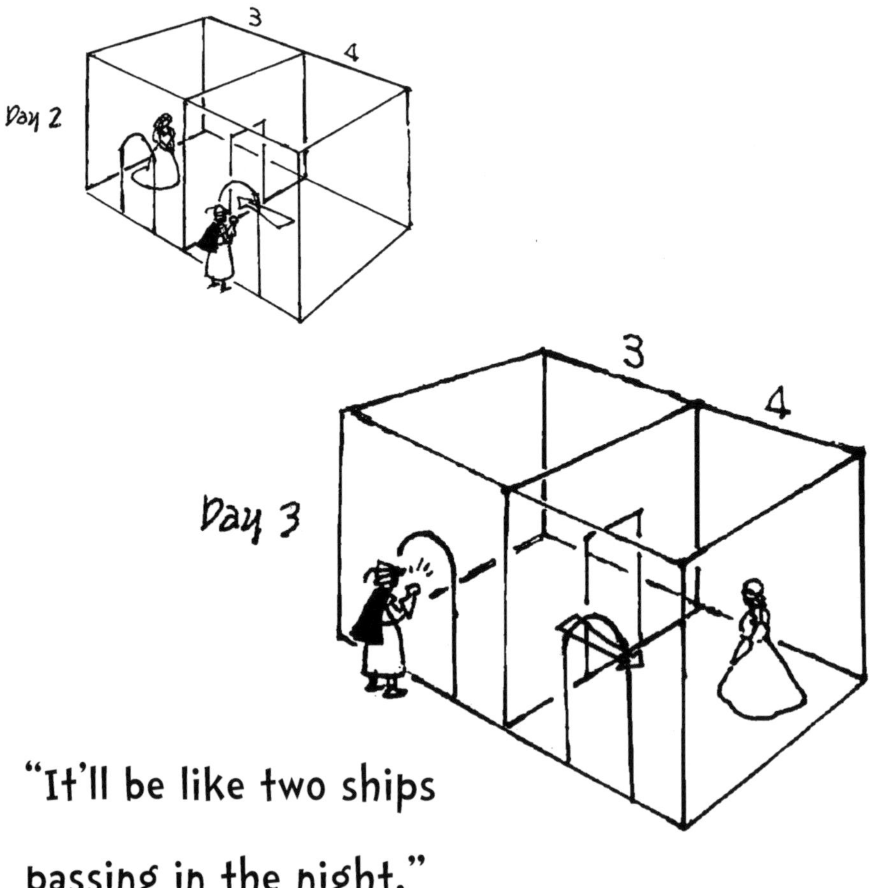

"It'll be like two ships
passing in the night,"
the prince moaned.

"Let's not lose heart.

We have four more days.

Why not continue to extend our wall?"

the girl was determined again.

"With the white wall,

you have knocked on rooms 5, 4, 3 and 2.

Extending with a black wall, how about you

continue with rooms 2, 3, 4 and 5?

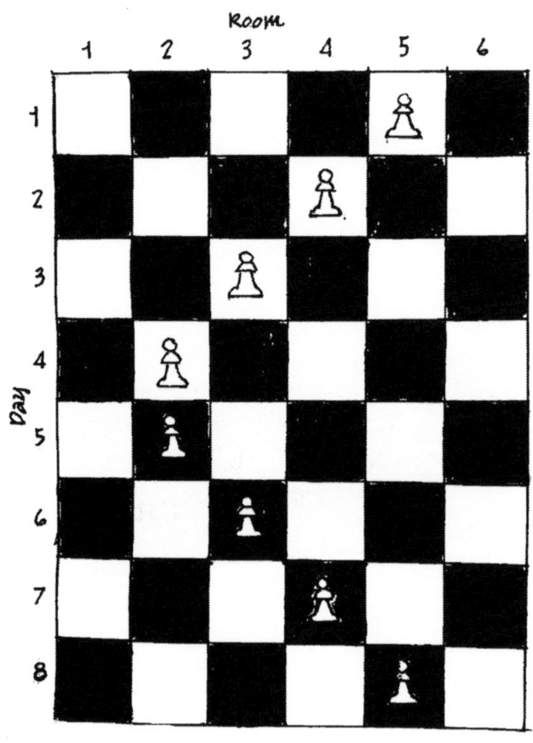

Now, let's see if she

can pass through that!"

The prince gasped.

He shivered with excitement.

The girl had represented the problem well.

She had changed his point of view.

She was leaving nothing to chance.

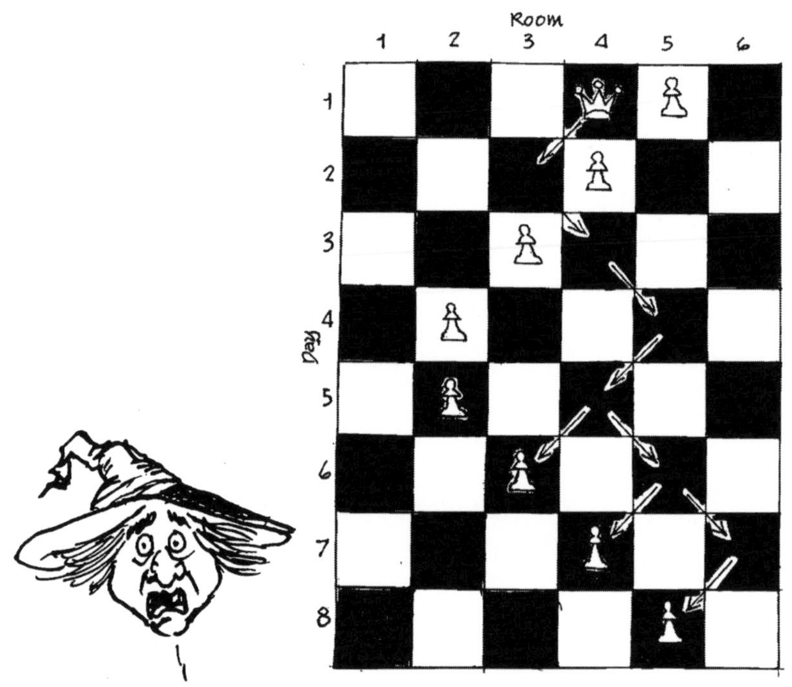

To be absolutely certain, the prince said,
"The 4-day wall on the white squares
connecting with a 4-day wall on the black
squares 'block' the princess' moves.
Knock the doors in the order
5-4-3-2-2-3-4-5 and the princess will be
rescued in at most eight days."

The prince thanked

the girl

profusely ...

and set off gallantly to carry out the plan.

And the girl was happy.

The End

The *I'm a Maths Star!* Guide to Problem-solving
Yeap Ban Har

What can we do when faced with a challenging Mathematics problem such as the one in *The Girl and the Princess*? How can we prepare ourselves to be good at problem-solving?

Here are six tips to keep in mind:

Tip #1:
Be like the girl: focus on the process, not the answer.
"We need a plan that will work whatever happens, ...
A foolproof plan that does not depend on luck and coincidence."

- We may **make a guess** but we should **check** that it will definitely work.

- We should make conclusions based on evidence rather than perception.

- **Don't rush** into giving an answer but have a plan that gives a solution that works.

Tip #2:
Be like the girl: monitor your thinking

- At the start, the prince suggested two ideas and each time the girl showed why the idea might not guarantee a correct solution.

- The girl personifies metacognition, our mind's ability to monitor itself.

- Metacognition helps us realise our ideas may not work and prompts us to seek better ones.

- Metacognition is important in becoming good problem solvers. We may start off like the prince—guessing possible answers.

- However, we must also end up like the girl—we must **leave nothing to chance and check that each step is a correct one.**

Tip #3:
Be like the girl: solve a simpler problem

The girl said softly, "It's usually good to work on something simpler. Let's try a row with only one room."

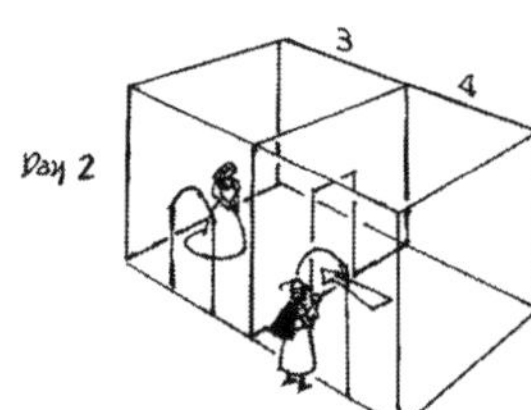

- **Simplify a problem** by changing the numbers.

- We can understand the problem better as we solve the simpler problems. We may also be able to see interesting patterns that emerge.

Tip #4:
Be like the girl: use various representations

"We need a new way to represent the situation," the girl said. "Let's use the chessboard as the grid."

- The girl suggested using a grid to organise important information, like where the princess might be on each day. She found the grid limiting and thought of using a more elegant representation: a chessboard with chess pieces.

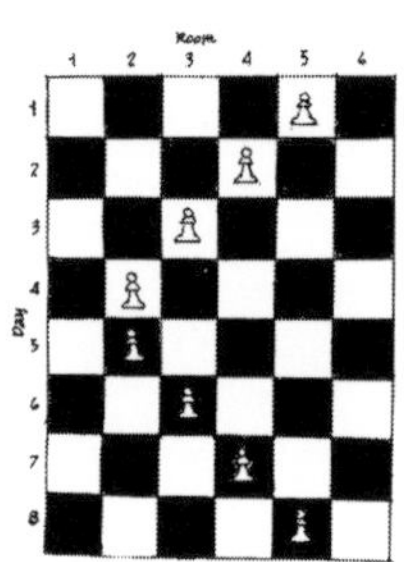

- When you **represent the Maths problem,** make sure each step follows from the previous and there are no loopholes.

- Always be inspired, like the girl when she thought of using the chessboard. **Drawing a diagram** is another good way of representing Maths problems.

Tip #5:

<h1 align="center">Be like the girl and the prince: persevere</h1>

- Don't give up easily. It took the girl and the prince quite some time to arrive at a solution. But what a reward when they finally did.

- And even if we don't arrive at a solution right away, it is fine. **The process itself makes us a better thinker.**

Tip #6:

<h1 align="center">Be like the prince: be willing to learn</h1>

- Always **be excited about learning** and enjoy the process!

For parents and teachers

Pólya's Model of Problem-solving

We would suggest that any attempt at Mathematical problem-solving requires a model to which the problem solver can refer, especially when one is unable to progress satisfactorily. A model helps in regulating a problem-solving attempt. We shall now describe the essential features of the problem-solving model proposed by George Pólya, an eminent Hungarian mathematician. The process can be summarised in the diagram below.

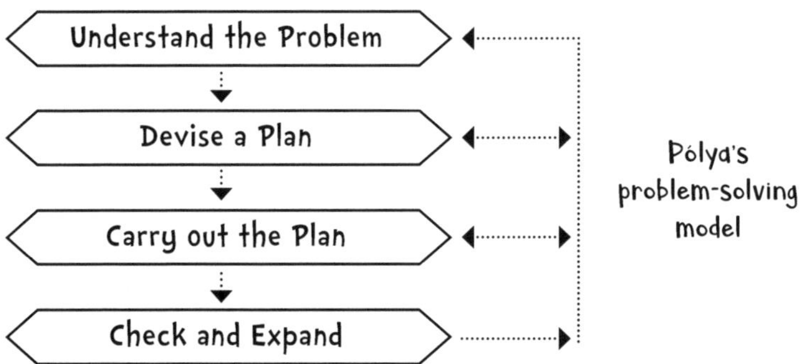

The model resembles a flowchart with four stages, Understand the Problem, Devise a Plan, Carry out the Plan, and Check and Expand, which are hierarchical but which allow back-flow. We shall explain the stages of the model and how the model works by applying it to our story of *The Girl and The Princess*.

Understand the Problem

Read aloud with the child. Pause to admire the nice illustrations. Take time to discuss and answer the child's questions, especially with regard to the arrangement of the rooms, the movement of the princess and the actions of the prince. The key parts of the problem to take time to understand would be:
- The princess has to move to an adjoining room every day
- The princess can move back to the previous room
- The prince can knock on the same door on consecutive days

It is common to see people staring blankly when they are stuck at a problem. It is important that something is done and the following describes how heuristics (meaning: serving to find out or discover, *Oxford English Dictionary*) can help start the problem-solving process.

Two useful heuristics to help understand the problem are to act it out and to consider a simpler problem (smaller numbers). Consider the problem when there are only 1, 2 or 3 rooms. Then one can use simple tokens to represent the rooms and the princess and act out the scenarios as in the story. Greater clarity will result from this worthwhile use of time

Devise a Plan / Carry out the Plan

A plan is vital to the success of any endeavour. It has been said, "He who fails to plan, plans to fail."

In the story, the girl planned to proceed with 5 rooms as she did with fewer rooms. However, she found that it had become too complicated. She then made a decision to change her plan. This resulted in a beautiful representation of the problem as one of the princess moving diagonally on a chessboard!

Check and Expand

It is good to look back at the solution and check it again. Check for various starting positions of the princess to see if the 5-4-3-2-2-3-4-5 process will catch/free her.

We are done for this problem but not done with the problem-solving process. It is a key feature of the model that the solver should try to 'expand' the problem even though it is solved. By expanding, we mean one of the following:
- find other solutions which are 'better' in the sense of elegance, succinctness, or with a wider applicability
- pose new problems

Indeed, we note that 5-4-3-2-2-3-4-5 is not the only solution. As long as we build two walls on white squares and black squares, we will succeed. Thus, another solution can be 2-3-4- 5-5-4-3-2.

We pose the following problems for the reader's consideration.
- What if there are 10 rooms? 20 rooms? How many days would the prince need in these cases?
- What if the princess was not allowed to change rooms for one night?

By thinking through the new problems, the child will reinforce his/her understanding of the original problem.

Scan this QR code to get additional practice on how to solve similar problems. Solutions are provided too.

Learn to solve difficult Maths problems using Maths heuristics, as identified by the Singapore Maths curriculum!

The *I'm a Maths Star!* series comprises challenging Maths puzzles, presented in story form. Puzzles are described and then solved, step-by-step, through an engaging storyline, and using various Maths heuristic techniques that are also taught as part of the world-renowned Singapore Maths curriculum. Through illustrations and an engaging storyline, children will learn Maths heuristics, and be inspired to persevere in understanding, representing, and solving fun and intriguing Maths problems!

To receive updates about children's titles from WS Education, go to: *https://www.worldscientific.com/page/newsletter/subscribe*, choose "Education", click on "Children's Books" and key in your email address.

Follow us @worldscientificedu on Instagram and @World Scientific Education on YouTube for our latest releases, videos and promotions.